Easy ISO

9001:2015

For All

(Employees & Employers)

Jahangir Asadi

Vancouver, BC CANADA

Published by: Silosa Consulting Group Inc.
Vancouver, BC **CANADA**
Email: Info@Silosa.ca
www.silosa.ca

Ordering Information:
Quantity sales. Special discounts are available on quantity purchases by universities, schools, corporations, associations, and others. For details, contact the "Sales Department" at the above mentioned email address.

Easy ISO 9001:2015 (for all employees and emplyers)/J.Asadi—1st ed.
ISBN: 978-1-990451-05-8

Contents

This book is dedicated to my first manager, **Mir Ahmad Sadat**, a Distinguished man of ISIRI, who has been my supporter and has encouraged me greatly and believed genuinely in my ability with honesty and self-confidence.

SILOSA Consulting Group (SCG)

Silosa Consulting Group (SCG) was established to provide outstanding consulting services of management system standards to individuals, groups, companies, and organizations all over the globe.

SCG is publishing an "EASY ISO" book series related to International Management System Standards to increase public knowledge in implementing these systems over their organizations.

SCG book publishing services and distribution services are connected to over 39,000 booksellers worldwide, including Apple, Amazon, Barnes & Noble, Indigo, Google Play Books, and many more.

We focus on quality, environmental, food safety and other management system standards.

SCG has enough experiences to help create new and effective programmes in different countries all over the world. For more detail, visit our website : http://silosa.ca and/or send your enquiery to the following email:

info@silosa.ca

About ISO

The International Organization for Standardization is an independent, non-governmental organization, the members of which are the standards organizations of the 165 member countries. It is the world's largest developer of voluntary international standards and it facilitates world trade by providing common standards among nations. More than twenty thousand standards have been set, covering everything from manufactured products and technology to food safety, agriculture, and healthcare.

Use of the standards aids in the creation of products and services that are safe, reliable, and of good quality. The standards help businesses increase productivity while minimizing errors and waste. By enabling products from different markets to be directly compared, they facilitate companies in entering new markets and assist in the development of global trade on a fair basis. The standards also serve to safeguard consumers and the end-users of products and services, ensuring that certified products conform to the minimum standards set internationally.

History

The organization began in the 1920s as the International Federation of the National Standardizing Associations (ISA). It was suspended in 1942 during World War II, but after the war ISA was approached by the recently formed United Nations Standards Coordinating Committee (UNSCC) with a proposal to form a new global standards body. In October 1946, ISA and UNSCC delegates from 25 countries met in London and agreed to join forces to create the new International Organization for Standardization. The new organization officially began operations in February 1947.

More information can be obtained :

www.ISO.org

Geneva, Switzerlan (Headquarter of ISO)

Definitions

ISO 9001 is defined as the international standard that specifies requirements for a quality management system (QMS). Organizations use the standard to demonstrate the ability to consistently provide products and services that meet customer and regulatory requirements.

Why is ISO 9001 important?

ISO 9001 aims to provide a practical and workable Quality Management System for improving and monitoring all areas of your business. ... Implementing an effective and robust ISO 9001 Quality Management System (QMS) will help you to focus on the important areas of your business and improve efficiency.

What is the meaning of ISO certified?

ISO certification is a seal of approval from a third party body that a company runs to one of the international standards developed and published by the International Organization for Standardization (ISO). ... ISO 9001 helps put your customers first.

Is ISO 9001 certification worth it?

Having an ISO certification is important for SMMs because of its ability to enact growth, profitability, and cost savings. The benefit of reducing waste also allows your workforce to be more efficient and establish ongoing QMS standards for improvement and sustainable customer success.

Products and/or Services Realization

1 How The Customer explained it

2 How the project leader understood it

3 How the analysit designed it

4 How the programmer wrote it

5 How the consultant described it

6 How the project was documented

Products and/or Services Realization

7 What operation installed

8 How the customer was billed

9 How it was supported

10 What the customer really needed

Customer satisfaction is the process of achieving goals and consistently delivering the value you promised. Customer delight, on the other hand, is the "wow" factor. It proves to the customer that you intimately understand their needs and can proactively anticipate ways to improve their situation.

Both are important aspects of customer success.

What Are the Benefits of Being ISO 9001 Certified?

1 Boost Employee Performance and Productivity

2 Define Your Company's Quality Control Processes

3 Reduce Waste and Improve Efficiency

4 Provide an Improved Customer Experience

5 Increase Confidence in your Manufacturing Business

1: Boost Employee Performance and Productivity

Engaged employees are motivated to implement processes that are put in place to ensure that problems are
quickly identified and resolved in a timely manner. Addi-
tionally, the consistent process audits through ISO 9001
can keep your employees focused while providing critical
feedback when your processes deviate from consistency.

2: Define Your Company's Quality Control Processes

A cornerstone component of ISO 9001 certification is establishing thorough business processes and defining responsibilities for quality control—and equally as important, relaying those specifications to employees. After all, 85% of employees are most motivated when internal communications are effective! Implementing ISO certification requirements provides valuable key performance metrics, such as on-time delivery, throughput, and overall equipment effectiveness, to accurately reflect your system's performance. These metrics will help you make more educated decisions to improve growth and profitability throughout your company.

3: Reduce Waste and Improve Efficiency

Earning the ISO 9001 certification enforces a continuous improvement strategy, so that you're always, by design, seeking ways to reduce waste and improve efficiency. Utilizing the requirements set forth in ISO 9001 will help to first identify areas of waste and then implement preventative measures to avoid wasteful situations. Streamlining your manufacturing operations from the ground up through ISO 9001 means every moving part is as effective as possible rather than discarded or unused.

4: Provide an Improved Customer Experience

The ISO 9001 certification process provides an enhanced customer service experience by not only identifying key priorities for customers, but also outlining processes to further optimize these priorities based on customer expectations and needs. Products with enhanced quality translate into reduced customer complaints and more satisfied consumers. The most successful manufacturers of today know that delivering a better product experience is what will keep customers coming back. Producing a product in a way that reduces waste and cuts costs means that you can bring more value to your customer, which further reinforces their loyalty to your business.

Increase Confidence in your Manufacturing Business

An ISO 9001 certification demonstrates to both customers and stakeholders that your manufacturing business has the capabilities of delivering high-quality products that meet all industry regulations and are delivered on time. This is crucial for your business, as your products should reflect the measures taken to create consistency and confidence in a global manufacturing market. An ISO 9001 certification also ensures that your business has all of the necessary tools, resources, and equipment for effectively producing your product or service.

INTRODUCTION TO QUALITY MANAGEMENT SYSTEMS

What is Quality ?

How would you describe what "Quality" means ?

Degree to which a set of inherent characteristics fulfils requirements

QUALITY DOES NOT OCCUR BY ACCIDENT

Identify, understand and agree customer requirements

Plan to achieve them

Measure, monitor & control processes/activities

REQUIRES A SYSTEM

Understanding the Customer's Requirements

Which is better, quality or quantity?

QUALITY DOES NOT OCCUR BY ACCIDENT

Identify, understand and agree

customer requirements

Plan to achieve them

Measure, monitor & control

processes/activities

REQUIRES A SYSTEM

What does PDCA stand for?
Please-Don't-Change-Anything...No! No! No!

Plan-Do-Check-Act
The Plan-Do-Check-Act (PDCA) Cycle is a four-step problem-solving iterative technique used to improve business processes

You can use this technique in all aspects of your life.

26

System to establish policy and objectives and to achieve those objectives

It means that:
You are in **Point A** in 2021 and would like to reach to **Point B** in 2025

Have you set your personal goal?
Goal setting is a powerful process for thinking about your ideal future, and for motivating yourself to turn your vision of this future into reality.

An objective is a statement which describes what an individual, team or organisation is hoping to achieve. ... Objectives are '**SMART**' if they are Specific, Measurable, Achievable, (sometimes agreed), Realistic (or relevant) and Time-bound, (or timely).

What is Quality Policy?

A Quality Policy is typically a brief statement that aligns with an organization's purpose, mission, and strategic direction. It provides a framework for quality objectives and includes a commitment to meet applicable requirements (ISO 9001, customer, statutory, or regulatory) as well as to continually improve.

Do you have a Quality Policy for your life?

Keep in mind that sometimes auditors like to test if employees are aware of the quality policy and if they understand it; internal communication of the quality policy statement is vital for ISO certification and is understood within your company.

Management system
to direct and control an
organisation
(Factory, house, life,..)
with regard to quality

Are you taking care of delivery times to
your customers?

PURPOSE OF ISO 9001

"ISO 9001 specifies the requirements for a quality management system that may be used for internal application by organisations, certification, or contractual purposes."

ISO 9001

Customer Request → *Customer Satisfaction*

INPUT → PROCESS → OUTPUT

ISO 9001 cover all activities between

customer request to customer satisfaction

QUALITY MANAGEMENT PRINCIPLES

- Customer focused organization
- Leadership
- Involvement of people
- Process approach
- System approach to management
- Factual approach to decision making
- Mutually beneficial supplier relationships
- Continual improvement

General

Identify processes, based on the following approch:

INPUT ●———➤ 𝕻𝕽𝕺𝕮𝕰𝕾𝕾 ●———➤ OUTPUT

Determine sequence and interaction,
Measure
Monitor and
Analyse

You can do it for the processes of your life

Documentation requirements

Documented procedures required by this standard and those needed by the organisation to control its processes.

The extent of the QMS documentation depends on the following:

a) size and type of the organisation

b) complexity and interaction of the processes

c) competence of personnel

Secret of Effective Management System Documentation

When you develop any documentation, verify it with the follwoing rules:

5 **W**'s   and

1 **H**

If your document can answer these 6 questions, then you have developed a completely effective document; no matter that it is a quality manual, procedure, SOP, work instruction,.......

Who, When, Where, Why, What and How

WHO   WHAT   WHERE   WHEN   WHY   HOW

Who	Customers, Employees, Suppliers, Competitors, Govenrment
What	Strategy (Corporate, Business Unit, Marketing Product)
Where	Markets, Facilities, Distribution, Outsourcing
When	Strategic Plan. Annual Plan, Proogram and Projct Management
Why	Leadership, Communities, Culture, Change management
How	Marketing, Operating Pian, Sales Force, Metrtics, Incentives

Quality Manual

Establish & maintain a manual including:
scope of QMS with details/justification
for any exclusions
procedures or reference to them
description of sequence and interaction
of processes included in QMS

What should a quality manual include?

As a Quality Manual is the same as a user manual of an equipment so the comparisons are as follow:

Use Manual of an Equipment	a Quality Manual
Introduction	Introduction of organization
Contents	Contents
a Detailed Photo of Parts	Organizational Chart
Description of Parts	Duties and Responsibilities
Meeting Needs	Meeting requirements of Standard

Don't forget that the "Standard" is a Question Book and the "Quality Manual" is an Answer Book, thus repeating the sentences of standards for each clause in the quality manual shows that the writer did not understand the concept of the minimum requirements for the quality management systems.

Control of Documents

QMS documents shall be controlled.

A documented procedure shall be established:

to approve documents prior to use

to review/update & re-approve as necessary

to identify changes and current revision status

to ensure relevant versions are available

to ensure legibility and identification

to control documents of external origin

to control obsolete documents

Don't be afraid of the above mentioned requirements, you can meet all of these needs via the following EASY solutions:

- Create an account in one of the online drives with enough capacity of the volume of your documentation, usually 100 Gb is enough.

- Create a file system directory for keeping your manual and procedures there and giving share access to authorized persons and assign a person as administration. Then they shall take care that it's always an updated version of documents are online and obsolute documents are out of access.

Control of Records

QMS records shall be maintained to provide evidence of conformance to requirements and effective QMS operation.

A documented procedure shall be established for :

Identification, storage, retrieval, protection, retention time and disposition of records.

Don't Afraid of the above mentiond requirements, you can meet all of these needs via the following EASY solutions:

Records are the result of implementation of a system, So they are the most important objective evidences that show the performance of the system.

You can do the same as using documents for controlling the records. For some paper items, please scan and upload in appropriate drive.

Management Review

Review QMS at planned intervals to ensure suitability and effectiveness.

Review input:
- results of audits
- customer feedback
- process performance & product conformance
- status of preventive & corrective actions
- actions from earlier management reviews
- changes that could affect the QMS
- recommendations for improvemen

Management Review

Outputs from management review shall include decisions and actions related to:

- Improvement of the QMS and its processes
- Improvement of product related to customer requirements
- Resource needs
- Results of management reviews shall be recorded

The system is working for you (the system is fully integrated along your processes and eases your operations).

You are working for the system (the system is beside your operations and looks as an additional burden.)

Here is a famous fable of an elephant and five blind men used as a metaphor. This picture conveys that different people will look at the standard in their own sub-jective-way. It is also used illustrate that different people in an organization will have different perceptions of risks and opportunities that the organization needs to address at any given time.

The Most Important Clause:

Preventive Actions

- Identify action to prevent potential nonconformities
- documented procedure
- determine potential problems and their causes
- evaluate need for action
- implement preventive action
- record results of action taken
- review action taken

Bibliography:

Anttila, J., & Jussila, K. (2017). ISO 9001:2015- a questionable reform. What should the implementing organisations understand and do? Total Quality Management and Business Excellence, 28(9-10), 1090-1105. https://doi.org/10.1080/14783363.2017.1309119.

Astrini, N. (2018). ISO 9001 and performance: a method review. Total Quality Management & Business Excellence, doi: 10.1080/14783363.2018.1524293.

Bou-Llusar, J. C., Escrig-Tena, A. B., Roca-Puig, V., & Beltra´n-Martı´n, I. (2005). To what extent do enablers explain results in the EFQM excellence model? International Journal of Quality & Reliability Management, 22(44), 337-353.

Chatzoglou, P., Chatzoudes, D., & Kipraios, N. (2015). The impact of ISO 9000 certification on firms' financial performance. International Journal of Operations and Production Management, 35(1), 145-174. https://doi.org/10.1108/IJOPM-07-2012-0387

Chiarini, A. (2017). Risk-based thinking according to ISO 9001:2015 standard and the risk sources European manufacturing SMEs intend to manage. The TQM Journal, 29(2), 310-323. https://doi.org/10.1108/TQM-04-2016-0038.

Domingues, J. P. T., Sampaio, P., & Arezes, P. M. (2016). Integrated management systems assessment: a maturity model proposal. Journal of Cleaner Production, 124, 164-174, doi: 10.1016/j.jclepro.2016.02.103

Gigante, N., & Ziantoni, S. (2015). L'edizione 2015 della norma ISO 9001, 2015. Retrieved from:https://www.accredia.it/app/uploads/2015/12/6050_5_L__700_edizione_2015_della_norma_ISO_9001___Arch__Gigante__Dr__Ziantoni.pdf

ISO (2015a). ISO 9001 - Quality management systems – requirements. Geneva: International Organization for Standardization.

ISO (2015). ISO Survey 2015 (online). Retrieved from: http//www.iso.org.

ISO (2018). ISO 19011 - Guidelines for auditing management systems quality management systems. Geneva: International Organization for Standardization.

ISO (2019). ISO 9000 Family - Quality Management. Retrieved from: https://www.iso.org/home.html.

Wilson, J. P., & Campbell, L. (2018). ISO 9001:2015: the evolution and convergence of quality management and knowledge management for competitive advantage. Total Quality Management and Business Excellence, pp. 1-16. https://doi.org/10.1080/ 14783363.2018.1445965

	# Easy ISO 9001:2015 Quality Management System
	# Easy ISO 13485:2016 QMS for Medical Devices
	# Easy ISO 14001:2015 Environmental Management System
	# Easy ISO 17025:2018 Testing and Calibration Laboratories

EASY ISO 22000:2018 For all Employees and Employers We Make it Easy JAHANGIR ASADI	# Easy ISO 22000:2018 Food Safety Management System
EASY ISO 26000:2010 For all Employees and Employers We Make it Easy JAHANGIR ASADI	# Easy ISO 26000:2010 Social Responsibility
EASY ISO 27001:2013 For all Employees and Employers We Make it Easy JAHANGIR ASADI	# Easy ISO 27001:2013 Information Security Management
EASY ISO 29001:2020 For all Employees and Employers We Make it Easy JAHANGIR ASADI	# Easy ISO 29001:2020 Petroleum, Petrochemical and Natural gas industries

	# Easy ISO 31000:2018 Risk Management System
	# Easy ISO 45001:2018 Occupational Health & Safety Management System
	# Easy ISO 50001:2018 Energy Management System
	**International Pocket Guide for** **HACCP**

www.ingramcontent.com/pod-product-compliance
Lightning Source LLC
Chambersburg PA
CBHW050125240326
41458CB00122B/1427